HOMESTEAD TOOLS AND SMALL ENGINE CARE

Keeping Saws, Mowers, Tillers, Generators, and Hand Tools Running

Hudson Briggs

Copyright

The information in this book is provided for general educational purposes. Power equipment is dangerous. Fuel is flammable and its vapor is explosive. Engine exhaust contains carbon monoxide, which is lethal and undetectable without an alarm. Electrical work carries risk of electrocution and fire. The manufacturer's manual for your specific equipment supersedes anything written here. The author and publisher disclaim all liability for any loss, injury, or damage arising from the use of this material.

Important Safety Notice

Never run an engine indoors, in a garage, in a basement, in a crawlspace, or near an open window or door. Carbon monoxide from a generator or any gasoline engine kills people every year, and a great many of those deaths involve a generator that was in an attached garage with the door open. That is not enough. Outdoors, well away from the building, exhaust pointed away, with working CO alarms inside the house.

Never plug a generator into a wall outlet. This practice - backfeeding - energizes the utility lines outside your house and can kill a lineworker or a neighbor. It is also illegal in most jurisdictions. Use a transfer switch installed by a licensed electrician, or run appliances directly on extension cords.

Disconnect the spark plug wire before working on any engine where the blade, chain, or driven part could move. On electric-start and battery equipment, remove the battery. Blades and chains can turn far enough under hand pressure to take fingers.

Handle fuel outdoors, with the engine cold and off. Gasoline vapor is heavier than air, travels along the ground, and finds pilot lights and water heaters. Never smoke around it. Never fuel a hot engine.

Wear the right protection. Eyes for anything spinning or cutting. Hearing for anything with an engine. Chainsaw chaps or cut-resistant pants and a helmet with a face screen for saw work, every time, no exceptions for short cuts.

Electrical work belongs to licensed electricians. Transfer switches, panel work, and permanent generator wiring all require permits in most places, and the inspection is the point, not the obstacle.

Have the fire extinguisher before you need it. An ABC-rated extinguisher in the shop, mounted by the door where you can reach it on the way past.

Contents

Introduction: The Machine Does Not Owe You Anything

I killed a string trimmer once by doing absolutely nothing wrong, which is to say I did nothing at all. Ran it in September, hung it on the wall, took it down in April, and it would not start. Pulled it until my shoulder hurt, sprayed starting fluid down the intake, pulled some more, and eventually took it to a shop where a man about seventy years old took the carburetor off, held it up to the light, and showed me a passage the diameter of a pencil lead completely closed with a hard green varnish. Sixty dollars. He was polite about it and I have never forgotten the lesson.

Here is what I did not understand at the time. Small engines almost never fail because a part wore out. They fail because of fuel that sat, oil that was never changed, or an air filter nobody looked at. Three things, and all three are prevented by habits that take less time per year than a single trip to the repair shop.

The mechanic told me something else that afternoon that turned out to be the most useful sentence anybody has said to me about equipment: most of what comes through this door is fuel. Not broken engines. Fuel.

What this book covers

The engines and tools an ordinary rural place actually runs - a chainsaw, a mower, a string trimmer, maybe a tiller, a log splitter, a pressure washer, a generator, and a wall of hand tools. It covers the maintenance that prevents failure, the diagnosis that finds a failure fast, and the repairs a reasonably handy person can do with hand tools in a shed.

It also covers what to hand to a shop. There is no shame in that line, and knowing where it falls saves money in both directions - you stop paying somebody sixty dollars to clean a carburetor you could have cleaned, and you stop destroying a two-hundred-dollar part trying to save forty.

What it does not cover

It is not a repair manual for your specific engine. Your manual has the torque specs, the clearances, the oil capacity, the plug gap, and the part numbers, and it wins any argument with this book. If you have lost it, nearly every manufacturer posts them free as PDFs by model number, and the model number is on a sticker or stamped tag on the engine.

It also does not cover automotive work, diesel tractors, or anything with a transmission you would need a shop press to open.

Three habits

Fuel discipline. Know how old the gas in every can and every tank is. This alone prevents most of what goes wrong.

Look at things. Two minutes of looking before you start a machine - oil level, air filter, chain tension, loose bolt, fuel line condition - catches nearly everything before it becomes expensive.

Write it down. A log book or a piece of masking tape on the machine with the date of the last oil change. Hours run, not months, is what maintenance intervals are actually built on, and nobody remembers.

About the numbers

Where this book gives a specification - an oil viscosity, a mix ratio, a fuel storage life - it names where it comes from, and manufacturers are cited where their guidance is the authority. Ratios are given both as ratios and in ounces per gallon, because ounces per gallon is what you actually need standing at the bench.

Part One - Fuel, Oil, And The Things That Actually Kill Engines

Chapter 1: Why Small Engines Die

Ask any small engine shop what comes through the door and you get the same answer everywhere. It is not worn rings and it is not burnt valves. It is four things, in this order.

Stale fuel

The overwhelming winner. Gasoline is not a stable product; it begins to oxidize and lose its light volatile fractions within weeks of leaving the refinery. What is left behind is heavier, harder to vaporize, and eventually deposits gum and varnish.

Oklahoma State extension's guidance on this is blunt: even straight gasoline should not be stored for more than sixty days without stabilizer, and for equipment stored longer than sixty days, draining the fuel is the usual recommendation.

Sixty days. Think about how that compares with your lawn mower, which sits from October to April with a half tank in it.

What the deposits do is close off the tiny passages in a carburetor. A small engine's main jet and idle circuit have orifices smaller than a pin. It does not take much varnish to close one, and once closed, no amount of pulling the rope will open it.

Ethanol, on top of stale fuel

Ethanol makes the storage problem worse rather than causing a separate one. It attracts water, it is a solvent that loosens deposits and old fuel line material, and in blends above ten percent it is hard on the rubber and plastic in equipment never designed for it.

Chapter 2 covers this in full, including a legal point most people do not know: E15 is prohibited by federal law in small nonroad engines.

Dirty air filter

An engine is an air pump. Restrict the air and it runs rich, fouls plugs, loses power, and washes oil off the cylinder walls with excess fuel. A filter costs a few dollars and takes ninety seconds to check.

Foam filters in particular need to be washed and re-oiled, and almost nobody does it. A dry foam filter passes dirt straight into the engine.

Old or absent oil

Four-stroke engines burn a little oil and leak a little, and a small engine has a very small sump - a walk-behind mower runs on fifteen to eighteen ounces, which is roughly a pint. A quart low is not possible; a pint low is catastrophic.

Small engines also run hot, dirty, and at constant high RPM, which is harder service than a car engine sees. Briggs & Stratton's interval for a walk-behind is fifty hours or annually, whichever comes first, and a hundred hours or annually for riding mowers.

Two-stroke engines have the opposite failure: running them on straight gasoline, with no oil mixed in, seizes them within minutes. Every year somebody fills a chainsaw from the mower can.

What almost never fails

Rings, valves, bearings, and crankshafts, on equipment that has been maintained. A well-kept small engine will outlast the machine it is bolted to. The corollary is that when a shop tells you a small engine has low compression, the question to ask is what killed it, because something did.

Chapter 2: Fuel, and Why Yours Is Probably the Problem

If you take one chapter out of this book, take this one. Fuel discipline prevents more repairs than every other maintenance task combined.

What ethanol actually does

Nearly all pump gasoline in the United States contains up to ten percent ethanol. Oklahoma State extension notes that manufacturers accept up to E10, and that ethanol is a solvent, absorbs water, and causes some swelling and change in rubber and elastomer parts - measurable, though in their testing not severe enough to fail.

The real damage comes from combination. Ethanol-blended fuel that sits picks up moisture, the light fractions evaporate, the ethanol scours old deposits loose from the tank and lines, and everything ends up in the carburetor. Phase separation - where the ethanol and water separate out into a corrosive layer at the bottom of the tank - takes a fair amount of water to occur, but a vented can in a humid shed over a wet summer has months to get there.

E15 is not legal in your equipment

This one surprises people. EPA approved E15 for model year 2001 and newer light-duty passenger vehicles and flex-fuel vehicles. It is prohibited in everything else - all motorcycles, all boat motors, all off-road vehicles, and specifically lawnmowers, chainsaws, leaf blowers, and small gasoline engines generally.

That is not a warranty preference. Federal law prohibits it, and the pump is required to carry a label saying so, headed "Attention." The label is small, the price is a few cents lower, and people fill their gas cans from it without reading a thing.

Look at the pump before you fill a can. E15 is sometimes sold as "Unleaded 88," which sounds like a premium grade and is not.

Ethanol-free gasoline

Many stations sell ethanol-free gasoline, often at the marina or in rural areas, at a premium of fifty cents to a dollar a gallon. For equipment that sits - a generator, a chainsaw, a tiller used twice a year - it is worth every penny. For a mower you run weekly all summer, ordinary E10 with stabilizer is fine.

Canned pre-mixed fuel, sold in quart and gallon containers, is ethanol-free, stabilized, and has a shelf life measured in years. It costs many times what pump gas costs, and it is still the right answer for a chainsaw you use six times a year and absolutely need to start when a tree is across the driveway.

The system that works

Buy small. A five-gallon can that lasts you four months is a five-gallon can of increasingly bad fuel. Buy the size you will actually use in a month.

Date every can. Masking tape and a marker on the side, the day you fill it. This is the single highest-value habit in this book. When the tape says more than sixty days ago, that fuel goes in a vehicle, not in equipment.

Stabilize everything at the pump. Add stabilizer to the can when you fill it, not when you decide to store the equipment, because the deterioration starts immediately. Follow the label rate. Stabilizer slows oxidation; it does not restore fuel that has already gone off.

Use metal or approved plastic cans, tightly sealed, stored somewhere cool and dark and never in the sun. Heat and light accelerate everything.

Never keep straight gasoline and two-stroke mix in identical cans. Color-code them, label them in large letters, and keep them apart. Running straight gas in a two-stroke destroys it in minutes.

Storing equipment for the season

Two approaches, and both work if you actually do them.

Run it dry. Shut off the fuel valve, if there is one, and let the engine run until it stops on its own. Then pull the rope a few times to clear the carburetor bowl. This is the better method for anything sitting more than a couple of months, and it is what I do with everything two-stroke.

Fill it and stabilize it. Fill the tank completely - a full tank has less air in it and therefore less moisture - with fresh, properly stabilized fuel, then run the engine five to ten minutes so the treated fuel is actually in the carburetor. This is the better method for four-strokes with vented steel tanks, where an empty tank rusts.

The mistake is doing neither, which is what almost everyone does.

For anything sitting more than about six months, drain the carburetor bowl as well. Most have a drain screw at the bottom. That bowl is where varnish forms.

Fuel line and filter

Fuel lines get hard, crack, and weep. On two-stroke equipment the lines are inside the tank and become brittle enough to crumble. On anything more than five years old, look at the lines when you service it, and replace them at the first sign of stiffness or cracking. Line is a few dollars a foot and the job takes twenty minutes. A cracked line lets air into the fuel system, and an engine drawing air runs lean, which is what burns pistons.

Inline fuel filters cost a couple of dollars and should be replaced annually on anything that has one.

Chapter 3: Mixing for Two-Strokes

Two-stroke engines have no oil sump. All lubrication comes from oil suspended in the fuel, which means the mix ratio is not a suggestion, it is the engine's oil supply.

Too little oil seizes the engine, usually within minutes and always permanently. Too much oil fouls plugs, carbons up the exhaust port and muffler screen, and makes it smoke and lose power - annoying but not fatal.

The ratios

Your manual specifies the ratio for your equipment. Most modern handheld two-stroke equipment - chainsaws, string trimmers, blowers, hedge trimmers - runs 50:1. Older equipment often calls for 40:1 or 32:1, and very old gear sometimes 25:1 or even 16:1.

Ratio	Oil per gallon	Per 2 gal	Per 2.5 gal	Per 5 gal
50:1	2.6 oz	5.1 oz	6.4 oz	12.8 oz
40:1	3.2 oz	6.4 oz	8.0 oz	16.0 oz
32:1	4.0 oz	8.0 oz	10.0 oz	20.0 oz
25:1	5.1 oz	10.2 oz	12.8 oz	25.6 oz
16:1	8.0 oz	16.0 oz	20.0 oz	32.0 oz

The arithmetic is simply 128 fluid ounces in a gallon divided by the ratio. At 50:1 that is 2.56 ounces, which everyone rounds to 2.6.

When in doubt, use the richer ratio. If you honestly cannot determine what an old machine wants, 40:1 will not hurt a 50:1 engine, and 50:1 will destroy a 32:1 engine.

The oil

Use oil made for air-cooled two-stroke engines. The specifications to look for on the bottle are ISO-L-EGD and JASO FD, which are the current high-performance ratings. Anything meeting those is fine, and the manufacturer's own branded oil is a safe default.

Do not use outboard motor oil, which is formulated for water-cooled engines that run far cooler. Do not use car oil. Do not use bar and chain oil.

Manufacturer-branded oils frequently include stabilizer, which is worth having.

How to mix

Do it outdoors, in a clean, clearly labeled can that never holds anything else.

Pour in about half the gasoline. Add all of the oil. Cap it and shake it thoroughly. Then add the rest of the gasoline and shake again. Adding oil to an empty can and then dumping fuel on it leaves oil stuck to the bottom.

Label the can in large letters with the ratio and the date. Then shake it before every fill, because mix separates over weeks.

Shelf life

Mixed fuel is worse than straight gasoline, not better. Two-stroke oil accelerates the breakdown.

Mix no more than you will use in about thirty days, and treat sixty days as the absolute outside limit even with stabilizer. Dispose of anything older by burning it in a vehicle you do not care much about, or by taking it to a household hazardous waste day.

The pattern that ruins saws is mixing five gallons in April because it seemed efficient, then running the last of it in September.

For equipment used a few times a year, buy canned pre-mixed fuel. The math works out in favor of it, and it will still start in three years.

Chapter 4: Oil for Four-Strokes

Mowers, tillers, generators, log splitters, and pressure washers mostly run four-stroke engines with a sump, and the whole discipline is checking the level and changing it on schedule.

Which oil

Briggs & Stratton's guidance covers most of the field, and the other manufacturers are close enough that following it will not get you in trouble.

Oil	Use
SAE 30	Warm temperatures; the most common small engine oil
SAE 10W-30	Varying temperatures, improved cold starting
Synthetic 5W-30	Best protection across all temperatures
SAE 5W-30	Very cold conditions
15W-50	Continuous heavy-duty use, such as commercial mowing

Two limits worth knowing. Below 40 degrees F, straight SAE 30 makes an engine hard to start. Above 80 degrees F, 10W-30 may increase oil consumption - which is not damage, but it means you check the level more often.

Synthetic 5W-30 or 15W-50 is approved across all temperature ranges, and for equipment that sits outdoors through hot and cold seasons, it is the simplest single answer.

Use a high-quality detergent oil rated for service SF, SG, SH, SJ or higher. Do not use special additives. The manufacturer's position on that is explicit, and a small engine sump is not the place for supplements.

How much, and how often

Capacities are small enough that guessing is dangerous. Briggs gives fifteen or eighteen ounces as typical for a walk-behind and forty-eight or sixty-four ounces for a riding mower. Your manual has the exact figure. Roughly a pint for a push mower and two quarts for a rider.

Change intervals: fifty hours or annually for walk-behind engines, a hundred hours or annually for riders, whichever comes first. New engines usually want a first change after the first five hours, and that break-in change matters more than any that follow it, because it carries out the metal from initial wear.

Check the level before every use. It takes ten seconds. Small sumps have no margin.

Doing the change

Warm the engine up first so the oil flows and carries the suspended dirt with it. Then shut it off and pull the plug wire.

Most walk-behind mowers have no drain plug and you tip them to drain out the fill tube. Tip toward the correct side - the manual says which, and it is usually with the air filter and carburetor up. Tipping the wrong way floods the carburetor and air filter with oil and gives you an afternoon of smoke and swearing.

An oil extractor pump that suctions out through the dipstick tube costs twenty or thirty dollars and eliminates all of that. It is the single best small purchase in this chapter.

Where there is a drain plug, position a pan, pull the plug, let it drain fully, reinstall the plug snugly without gorilla torque - most are into aluminum, and stripped threads in an aluminum crankcase are a bad day.

Refill to the mark and no further. Overfilling causes as much trouble as running low - the crankshaft whips the oil into foam, foam does not carry a load, and the engine smokes and pushes oil out the breather. Check on level ground, and follow your manual on whether to screw the dipstick in or just rest it.

Take used oil to any auto parts store or municipal recycling center. Most take it free.

Oil filters and air-cooled reality

Most small engines have no oil filter at all, which is exactly why the change interval is short. Some larger engines do, and those get a new filter with every change.

These are air-cooled engines. Cooling fins clogged with grass clippings and dust cause overheating, and overheating destroys oil. When you change the oil, blow or brush the fins clean and clear the screen over the flywheel. On a mower that has been used hard in dry conditions, skipping this is how engines get cooked - a packed cooling shroud has nowhere to send the heat.

Part Two - Maintenance That Matters

Chapter 5: Air, Spark, and the Twenty-Minute Service

Once a season, per machine, twenty minutes. This is the whole program for most equipment.

Air filter

Pull it and look at it in the light. If you cannot see light through it, it is done.

Paper filters get replaced, not cleaned. Tapping one out gently to knock off loose debris is fine. Blowing compressed air through one blows holes in the media and defeats the purpose. They cost a few dollars.

Foam filters get washed in warm soapy water, squeezed - not wrung, which tears them - rinsed, dried completely, then lightly oiled with clean engine oil and squeezed out so they are damp, not dripping. The oil is what actually catches the dust. A dry foam filter is a hole with a sponge in it.

Dual-element filters have a foam pre-cleaner wrapped around a paper element. Wash and oil the foam, replace the paper.

Check filters more often in dusty work - tilling, mowing in drought, running a splitter in a gravel yard. Every few hours in bad conditions is not excessive.

Spark plug

Pull it once a year with the correct socket. Look at the electrode.

Light tan or gray is a healthy engine. Black and dry sooty means it is running rich, which usually means a restricted air filter or a carburetor problem. Black and wet with oil means oil is getting into the combustion chamber, which on a four-stroke is worn rings or valve guides and on a two-stroke is usually too much oil in the mix. White or blistered means it is running lean or too hot, which is worth chasing down because lean kills pistons.

Gap it with a feeler gauge to the manual's spec - commonly .030 inch on small engines, but confirm yours. Clean a serviceable plug with a wire brush, or just replace it, because plugs cost a few dollars and a marginal plug causes hard starting that people chase for weeks.

Thread it in by hand first, every time. Cross-threading an aluminum head is an expensive minute.

Fasteners, blades, and the walk-around

Every engine vibrates and every vibrating machine loosens its own fasteners. Once a season, go over the machine with a wrench - engine mounting bolts, handle hardware, deck bolts, muffler bolts. A loose muffler bolt turns into a broken exhaust stud, which is a much larger job.

Look at the fuel lines for cracks and hardening. Look at the primer bulb, which cracks and stops working. Look at the recoil starter rope for fraying, because it always breaks in the middle of a job.

Grease whatever has grease fittings, per the manual.

Sparks and screens

Small engines have a spark arrestor screen in the muffler, which is a legal requirement on public land in many places and a fire prevention measure everywhere. It carbons up and chokes the exhaust, causing power loss that people blame on the carburetor.

Pull it once a season and clean it with a wire brush, or burn the carbon off carefully with a torch. Replace it if it is broken or perforated.

What a service actually looks like

For a walk-behind mower, start to finish, on a bench or on the ground with the plug wire off.

Check and drain the old oil, or extract it through the dipstick tube. Refill to the mark with the correct viscosity. Pull the air filter and either replace the paper element or wash, dry, and oil the foam. Pull the plug, read it, gap it or replace it, and thread it back by hand. Remove and sharpen the blade, balance it, and reinstall it right side down at the correct torque. Scrape the deck. Brush the cooling fins and the flywheel screen clear. Go around the machine with a wrench on every visible bolt. Look at the fuel line, the primer bulb, and the starter rope. Wipe it down and note the date on a piece of tape.

That is twenty to thirty minutes, once a year, and it is essentially the entire difference between a mower that lasts five years and one that lasts twenty.

The same sequence, with the machine-specific items swapped in, covers a tiller, a generator, a pressure washer, or a rider. The order matters only in that draining warm oil first gives you something to do while it drains.

Chapter 6: The Carburetor

More small engine problems live here than anywhere else, and understanding roughly how it works turns a mystery into a checklist.

What it does

A carburetor's job is to mix fuel with incoming air in the right proportion. Air rushing through a narrowing in the throat creates low pressure, which pulls fuel through a jet into the airstream. That is the whole principle.

Everything else is refinement. A float and needle valve keep the fuel bowl at a constant level. A separate idle circuit feeds the engine at low throttle when there is not enough airflow to pull fuel through the main jet. A choke plate restricts incoming air for a cold start, richening the mix. A primer bulb squirts raw fuel in to get things going.

The passages that carry fuel through all of that are very small. Some are the diameter of a heavy sewing needle.

Why they fail

Almost always deposits, and almost always from fuel that sat. Evaporating gasoline leaves gum, gum hardens into varnish, varnish closes the small passages, and the idle circuit - smallest of all - goes first.

That gives the classic pattern: the engine starts and runs on choke or at full throttle but dies at idle. That is a plugged idle circuit almost every time.

The other failure is a stuck float or a needle valve held open by debris, which floods the engine, or one stuck shut, which starves it.

Cleaning one

This is within reach of a patient person with hand tools, and it is the repair most likely to save you real money.

Work outdoors or with real ventilation. Carburetor cleaner is nasty and gasoline vapor is explosive. Wear eye protection - spraying cleaner through a passage will send it straight back at your face.

Photograph everything as you go. Especially the linkage, the springs, and how the governor spring attaches, because those go back exactly one way and no diagram in the world will help you remember which.

Shut off the fuel, drain what you can, and remove the carburetor. Take the bowl off, note where the float pivots, remove the float and needle. Remove the main jet, which on a simple engine is often inside the bolt that holds the bowl on and has a cross-drilled hole in the side that also needs to be clear.

Spray cleaner through every passage until it comes out the other end freely, and use the little red straw. Run a strand of fine wire or a bristle from a wire brush through the jets - a torch tip cleaner is the right tool. Do not drill a jet. Enlarging an orifice permanently changes the fuel mixture and there is no going back.

Blow everything dry with compressed air if you have it.

Replace the bowl gasket and the needle if the kit includes one. Rebuild kits cost ten to twenty dollars and are cheap insurance while you have it apart.

Reassemble using your photographs, install, and set the float level per the manual if it is adjustable.

Rebuild or replace

Aftermarket replacement carburetors for common engines have gotten inexpensive - often fifteen to forty dollars - and for a machine of modest value, replacing beats an afternoon of cleaning. Quality varies and the good ones are usually fine.

Rebuild when the engine is worth it, when the carburetor is unusual, or when you want to know how to do it. Replace when it is a common engine and your time is worth more than the part.

Adjustment screws

Some carburetors have adjustable high and low speed mixture screws, particularly on chainsaws and trimmers, and many modern ones are sealed or limited by law.

If yours adjusts, the base setting is usually about one and a half turns out from lightly seated, then fine-tuned with the engine warm. Turn the low speed screw for a clean idle and a crisp response to throttle, and the high speed screw so the engine four-strokes slightly at full throttle without load and cleans up under load.

Running a two-stroke lean at high speed will destroy it. If you are uncertain, err rich. A saw that smokes a little is running. A saw set lean is running out of piston.

Chapter 7: When It Will Not Start

An engine needs four things: fuel, air, spark, and compression, arriving at the right time. Every no-start is one of those four. Work them in order and you will find it, usually in ten minutes.

Start with the stupid things

Check the fuel level, and check that it is the right fuel in the right machine. Check the on-off switch. Check the fuel shutoff valve. Check the safety interlocks, which are on everything now - the bail on a mower handle, the seat switch and blade-engagement switch on a rider, the chain brake position on a saw. Check that the choke is where it should be.

I would guess a third of no-starts are on this list, and every one of them is embarrassing to find after an hour.

Spark

Pull the plug, reconnect the plug wire, lay the plug's threaded body firmly against bare metal on the engine, and pull the rope while watching the gap. You want a strong blue spark. Keep your fingers off it and keep the plug hole away from you, because it will spit fuel vapor.

An inline spark tester costs a few dollars and is safer and easier to see.

No spark points to a bad plug - try a new one first, since it is cheap - a disconnected or shorted kill wire, a bad ignition coil, or a sheared flywheel key. That last one is common on mowers that hit a rock or a stump: the soft key that indexes the flywheel to the crankshaft shears to protect the engine, the timing shifts, and it will not start. It is an inexpensive part and a moderately annoying job.

Fuel

Confirm fuel is reaching the cylinder. Pull the plug and look at it after several pulls - if it is wet with fuel, fuel is arriving. Dry means it is not.

If it is dry, work backward: fuel in the tank, valve open, filter not clogged, line not cracked or blocked, carburetor not varnished shut.

A useful test is a small squirt of starting fluid into the carburetor throat. If it fires for a second or two and dies, you have spark and compression, and your problem is fuel delivery - meaning the carburetor. If it does not fire at all, look at spark and compression instead.

Use starting fluid sparingly. It is a diagnostic, not a starting method. It has no lubricating properties, and habitual use scuffs cylinders - and on a two-stroke it does real harm.

Compression

Pull the rope slowly. You should feel a clear, distinct resistance as the piston comes up on compression. If it spins freely with no resistance anywhere, something serious is wrong - a broken rod, a hole in the piston, a valve stuck open, or a blown head gasket.

A compression gauge that threads into the plug hole costs twenty or thirty dollars. Typical healthy small engines read somewhere around 90 to 120 psi, and your manual gives the real figure. Note that many modern engines have a compression release built into the camshaft to make starting easier, which makes a cranking reading lower than you would expect.

Low compression on an engine that was maintained usually traces back to overheating, running lean, or oil starvation, which is why the earlier chapters matter.

Runs and then dies

A distinct problem with its own list.

Dies after a few minutes and restarts once cool: an ignition coil breaking down as it heats up, which is common, or a blocked fuel tank vent creating a vacuum. Loosen the fuel cap and see if it runs - that one takes five seconds to test.

Dies when you tilt it or turn: fuel pickup problem or a low tank.

Runs only on choke: lean condition. Plugged idle circuit, air leak at a cracked line or intake gasket, or a partly blocked jet.

Surges up and down at idle: usually a partly plugged idle circuit, sometimes a governor issue.

Loses power under load: fuel restriction, a plugged spark arrestor screen, or a dull cutting attachment making the engine work harder than it should.

Part Three - The Machines

Chapter 8: Chainsaws

A chainsaw is the most dangerous tool most people own and the one most improved by simple maintenance. A sharp chain is safer than a dull one, because a dull chain has to be forced and forcing is what causes accidents.

The chain

Sharpen far more often than you think. A working rule is every tank of fuel, and immediately any time the chain touches dirt, gravel, or a nail, which dulls it in one cut.

You will know it is dull when the saw makes fine powder instead of chips, when it wanders instead of cutting straight, and when you have to lean on it. A sharp chain pulls itself into the wood under the weight of the saw and throws coarse chips.

Round file, of the diameter your chain calls for. Oregon and the other manufacturers publish filing specifications by chain type, and the chain package and your manual carry the numbers - file size, top plate angle, and depth gauge setting. Get them from your specific chain rather than a general rule, because they vary meaningfully between chain types and using the wrong file diameter reshapes the cutter incorrectly.

The technique: clamp the bar in a vise or wedge the saw so it will not move. File each cutter from the inside out, in the direction the cutter faces, holding the file angle steady and using the file guide marks. The file only cuts on the push stroke - lift it on the return. Count your strokes and give every cutter the same number, so they stay the same length. Do all the cutters facing one direction, then turn around and do the others.

Uneven cutters make the saw cut in an arc. If some are much shorter, file them all back to match the shortest.

Depth gauges - the rakers, the small tabs ahead of each cutter - control how deep the cutter bites. As you file the cutters back, the depth gauges become proportionally too tall and the saw stops cutting well even though the edges are sharp. Check them with a depth gauge tool every few sharpenings and file the tops flat with a flat file to your chain's specification, then round the front edge back to shape. Take too much off and the saw grabs and kicks; leave them too tall and it will not cut.

Once a chain has been filed back to where the cutters are very short, replace it. Chains are consumables. Running two or three in rotation means one is always sharp.

Bar and drive

Flip the bar every time you change the chain so it wears evenly. Clean the bar groove with a screwdriver or a bar groove tool - packed sawdust and oil starve the chain of lubrication. Clear the oiler hole in the bar and in the saw body.

Check the rails for burrs and dress them flat with a file. Check for wear that lets the chain lean, because a worn bar makes the saw cut crooked no matter how well you file.

Use bar and chain oil, not used motor oil, which is thin, throws off the chain, contains combustion acids and metal particles, and is a genuine health hazard. A properly working saw uses roughly a tank of bar oil per tank of fuel, so fill both every time. If the bar oil level is not dropping, the oiler is blocked and the bar will be ruined shortly.

Chain tension: a cold chain should sit snug against the bottom of the bar, pulled down a little by hand and snapping back, and still turn freely by hand with a glove on. Chains loosen as they heat up and tighten as they cool, so check tension when cold and re-check after the first few minutes of cutting. A new chain stretches noticeably during its first hour and needs tightening several times.

Never tension a hot chain to snug, because it will shrink as it cools and can bind or damage the crankshaft bearings.

Safety systems

Check these every time and do not defeat any of them.

The chain brake must stop the chain instantly when the front handguard is pushed forward. Test it with the saw running at idle: engage the brake, then briefly increase throttle and confirm the chain does not move. If it moves, the saw is out of service until it is fixed.

The throttle interlock should prevent the throttle from opening unless the rear handle is properly gripped.

The chain catcher, the small pin under the saw, catches a thrown chain. Replace it if it is broken.

The anti-vibration mounts should be intact and not cracked or oil-soaked.

Wear the gear. Chaps or cut-resistant pants, a helmet with a face screen, hearing protection, gloves, and boots. Chaps work by pulling long fibers into the sprocket and jamming the saw, and they have saved a great many legs. Carry a tourniquet and a phone, and tell someone where you are.

Chapter 9: Mowers and String Trimmers

Mower blades

A sharp blade cuts grass. A dull blade tears it, leaving a ragged white-tipped lawn that browns off and invites disease, and it makes the engine work harder for a worse result.

Sharpen two or three times a season for an ordinary yard, more if you mow sandy ground.

Disconnect the plug wire. Tip the mower the correct direction - carburetor and air filter up - or better yet remove the blade with the deck on its wheels. Mark which side of the blade faces down before you remove it, because installing a blade upside down is a mistake people make and it does not cut at all.

The blade bolt is usually very tight; a breaker bar and a block of wood wedged against the blade to stop it turning will do it.

Sharpen with a file, an angle grinder, or a bench grinder, following the existing bevel, which is around thirty degrees on most blades. You are not making a knife. A mower blade edge should be about as sharp as a butter knife - sharp enough to cut grass, blunt enough not to chip when it hits a stone. A razor edge on a mower blade folds over in ten minutes.

Keep it cool if you are grinding. Blueing the steel means you have drawn the temper and the edge will not hold.

Balance the blade before it goes back on. Hang it on a nail through the center hole, or use a cone balancer. If one end drops, take a little more metal off that end and check again. An unbalanced blade at three thousand RPM vibrates hard enough to destroy crankshaft bearings, and that is an engine-replacement expense from a five-minute shortcut.

Replace a blade that is cracked, deeply nicked, worn thin, or has a rounded-off trailing edge. Blades are cheap and a blade that breaks at speed becomes a projectile.

The rest of the mower

Scrape the underside of the deck. Packed clippings hold moisture, rust the deck out, and disrupt the airflow that lifts the grass into the blade - a caked deck cuts noticeably worse.

Clean the cooling fins and flywheel screen as covered in Chapter 4. On a rider, this is the leading cause of overheated engines.

Check tire pressures on a rider, because uneven pressure gives an uneven cut and people spend hours leveling a deck that was fine. Set all four to the manual's figure before you touch a single adjuster.

Then level the deck properly, on a flat concrete floor, with the blades turned so they point front-to-back. Measure from the floor to the tip of each blade. Side to side should be equal or within about an eighth of an inch. Front to back, most manufacturers want the front tip a quarter inch or so lower than the rear, which is called rake and it is what keeps the deck from cutting the same grass twice. An unlevel deck is the cause of nearly every complaint about stepped or streaked cutting.

Grease the fittings. Check belts for glazing, cracking, and fraying, and replace one that is shiny on the sides, since a glazed belt slips under load and will not drive the deck through heavy grass.

Self-propelled walk-behinds have a drive cable, a belt, and a pair of drive gears in the rear wheels. When one stops pulling, the usual order is a stretched cable first - most have an adjuster at the handle - then a worn belt, then the plastic drive gears inside the wheels, which strip their teeth and are inexpensive to replace. Pull the wheels once a season, clean the packed grass out, and look at the gear teeth.

On riders, keep the battery charged over winter with a maintainer, and clean the terminals. A lead-acid battery left discharged through a cold winter sulfates and does not come back.

String trimmers

The head is the main service item. Wind new line neatly in the direction marked on the spool, keep it in even layers, and do not overfill, which causes jamming and welding.

Use the diameter of line your machine specifies. Heavier line than the engine can drive overloads it and burns clutches.

Line that has dried out gets brittle and breaks constantly. Storing spools in a sealed container with a little water in it restores flexibility, and that trick works well enough to be worth doing.

Check the shaft and gearbox. Straight-shaft trimmers have a solid drive shaft that needs occasional greasing at the gearbox; curved-shaft models have a flexible cable that should be pulled and greased once a season.

Clear the debris shield and keep the cutting guard blade in place, since that is what trims the line to the right length.

Two-stroke trimmers live and die by the fuel rules in Chapters 2 and 3. More trimmers are killed by mix problems than by anything mechanical.

Chapter 10: Generators

A generator is the one machine on the place that people buy, put in a shed, and never touch until the night they desperately need it. That is the worst possible ownership pattern.

Carbon monoxide, first

More people are killed by generators than by any other tool discussed in this book, and the cause is almost always carbon monoxide.

Run it outdoors only. Not in a garage with the door open. Not in a carport. Not in a shed, a breezeway, a basement, or a crawlspace. Outdoors, at least twenty feet from the house, with the exhaust pointed away from doors, windows, and vents.

Have working carbon monoxide alarms inside the house, on every level, tested. CO is odorless, colorless, and causes confusion before it causes collapse - meaning the person being poisoned frequently cannot recognize what is happening or act on it.

Many newer generators have CO shutoff sensors. They are a good backup and not a license to move it closer.

Never backfeed

Plugging a generator into a wall outlet with a double-male cord - a "suicide cord," which tells you what you need to know - sends power backward through your panel and out onto the utility lines, where it can kill a lineworker restoring your service.

It also bypasses every overcurrent protection in the house and is a fire risk, and it is illegal virtually everywhere.

The two legitimate options are a transfer switch or interlock kit, installed by a licensed electrician with a permit, which physically prevents the generator and the utility from being connected at the same time; or extension cords running directly from the generator to individual appliances, using heavy outdoor-rated cords of adequate gauge.

Sizing

Watts are the unit. Running watts are what a device draws continuously; starting watts are the surge an electric motor pulls for a second or two at startup, often two to three times its running draw. Your generator has to cover the total running load plus the largest single starting surge.

Load	Running watts	Starting watts
Refrigerator	600 - 800	1,800 - 2,400
Chest freezer	500 - 700	1,500 - 2,100
Well pump, 1/2 hp	1,000	2,000 - 3,000
Sump pump	800 - 1,000	1,600 - 3,000
Furnace blower	600 - 800	1,500 - 2,400
Lights, LED, whole house	100 - 300	-
Microwave	1,000 - 1,500	-
Window AC, 10,000 BTU	1,200	1,800 - 3,600

Add up what you actually need running at once, then add the largest starting surge, then add a margin. A 5,000-watt generator covers a refrigerator, a freezer, a well pump, lights, and a furnace blower for most households, provided the well pump and the refrigerator do not start at the same instant.

Watts equal volts times amps, so a 5,000-watt generator at 120 volts can deliver about 41 amps, which is the number that tells you what cords and breakers you need.

Do not run a generator at its maximum rating continuously. Most are rated for a lower continuous output than their peak, and running at eighty percent or less is easier on the machine and on your fuel.

Keeping it ready

This is the part people skip.

Exercise it monthly. Start it, run it for twenty to thirty minutes under some real load - a couple of space heaters or shop lights - and shut it down. Running under load keeps the carburetor clear, warms the oil, and drives moisture out of the engine and the windings. Running with no load at all does much less good.

Store it dry or store it stabilized. Chapter 2's rules apply with force here, because a generator sits longer than anything else you own. My preference is to shut the fuel valve and run the carburetor dry after every use, and to keep the tank empty.

Change the oil on schedule, including during extended outages. Small generators frequently want an oil change every fifty to a hundred hours, which during a long outage is a matter of days, not years. Keep oil and a filter on hand, because after a hurricane the parts store is closed.

Keep spares: oil, an air filter, spark plugs, and a length of fuel line. During a regional outage none of these are available.

Store fuel safely and legally. Gasoline in approved cans, cool and dark, dated and stabilized, and not in an attached garage in quantity. Check your local fire code, which limits how much you may store.

Test the cords and the transfer switch during your monthly exercise. That is the point of the exercise.

Running one during an outage

Set it up before dark and before the weather gets bad, on level ground, out of rain and snow but with real ventilation - an open canopy or a purpose-built generator cover, not a shed.

Start it with no load, let it warm up, then add loads one at a time, largest motor first, so you do not stack starting surges.

Refuel with the engine off and cool. Gasoline spilled on a hot muffler ignites. This is the second most common generator injury after CO.

Keep a log of hours run so you know when the oil change is due.

Chapter 11: Tillers, Splitters, Pumps, and Pressure Washers

Rototillers

Everything in Part One applies to the engine. The tiller-specific service is the transmission and the tines.

Check the gearbox oil per the manual - most use a heavy gear oil and most owners never look at it. Grease the fittings. Look at the tines for wear; when they are rounded off and shortened, the tiller stops digging and starts skating, and people blame the engine.

The belt and clutch are wear items. A glazed or cracked belt slips under load.

After the last use of the season, clean the tines completely and coat them with oil, because caked soil holds moisture and pits them badly over a winter.

Log splitters

Hydraulics rather than engine work. Check the hydraulic fluid level with the ram fully retracted and the machine level, and change the fluid and filter on the manual's schedule, which is usually annual or by hours.

Inspect hoses for cracking, bulging, chafing, and weeping at the fittings. Never check for a hydraulic leak with your hand. High-pressure fluid injection through the skin is a surgical emergency that people underestimate because the entry wound looks trivial. Use a piece of cardboard, and wear eye protection.

Keep the wedge sharp with a file and keep the beam clean and lightly oiled.

Operate with one person only. Two-person splitting - one feeding, one on the lever - accounts for a large share of amputations.

Water pumps

Transfer pumps and pressure pumps live in Chapter 2's world plus one extra rule that overrides everything else: never run a centrifugal pump dry. The seal is water-cooled and it takes very little running dry to destroy it. Prime it fully before starting.

Drain pumps completely before freezing weather. A frozen pump housing cracks, and the housing is most of the value of the pump.

Screen the intake, and keep the suction hose from collapsing, since a collapsed suction line will not prime.

Pressure washers

The pump is the expensive part and it is killed two ways: running dry and freezing.

Always connect and turn on the water supply and squeeze the trigger to purge air before starting the engine. Never let it run more than a couple of minutes with the trigger closed, because bypassed water heats up fast and cooks the seals.

Use pump saver - the antifreeze-and-lubricant product sold in aerosol cans - before winter storage. It costs a few dollars and it is the difference between a pump that works next spring and one that does not.

Clean the inlet screen and the nozzles. A clogged nozzle causes pressure spikes. Nozzle tips wear out and a worn tip gives poor pressure that people blame on the pump.

Part Four - Hand Tools And The Shop

Chapter 12: Sharpening Everything Else

A dull tool is more dangerous than a sharp one, because dull tools require force and force is what slips.

Axes and hatchets

An axe is a wedge, not a knife, and the shape of the wedge matters more than the keenness of the edge.

Start with a file if the edge is damaged, working from the edge toward the head with a bastard mill file, following the existing convex bevel. Do not create a flat bevel, and do not thin the edge to a knife profile - a thin axe edge chips and sticks in the wood rather than splitting it out.

Move to a coarse then fine sharpening stone, working in circles, keeping the convex shape. Finish with a strop or fine stone until you can see no light on the edge.

Splitting mauls want a duller, blunter profile than felling axes. A maul is a wedge on a stick and does not need to slice at all.

Keep the head cool if you use a grinder. Blue steel is ruined steel.

Shovels and hoes

Both work better sharp and almost nobody sharpens them, which is a shame because it takes two minutes with a file and roughly halves the work.

Clamp the tool, file the front face of the blade at about a forty-five degree angle, working the full edge. You are not making it razor sharp - you are restoring a working edge that will slice roots and sod instead of mashing them.

Round-point shovels get sharpened on the leading curve. Hoes get sharpened on the leading edge, on the side facing away from you as you pull.

Pruners and loppers

Bypass pruners have one cutting blade and one hook, and only the cutting blade gets sharpened, on the outside bevel only. Sharpening the inside face destroys the fit between the blades and the tool will never cut cleanly again.

Use a small diamond file or a carbide sharpener, following the existing bevel, with light strokes. A few passes is usually enough.

Clean sap off the blades with alcohol or a solvent, and take pruners apart once a year to clean and oil the pivot. Adjust the pivot bolt so the blades are snug without binding.

Disinfect pruners between trees when you are cutting diseased wood - a wipe with seventy percent alcohol is the standard practice and it prevents you from carrying disease down a row.

Anvil pruners crush rather than slice, which is fine for dead wood and bad for live stems.

Knives

The most-used tool on any place and the one worth learning to sharpen properly.

Hold a consistent angle - about twenty degrees per side for a general working knife, less for something meant for fine slicing. Work one side until you can feel a burr along the full length of the opposite side, then do the other side, then alternate lighter strokes to remove the burr. Finish on a strop.

The single mistake beginners make is inconsistent angle, which rounds the edge instead of sharpening it. A cheap angle guide fixes it, and so does practice.

A honing steel does not sharpen. It realigns an edge that has rolled over, and it should be used often, with light strokes.

Saw blades and chisels

Hand saw blades can be filed and set, and it is a real skill worth learning if you own good saws. Modern hardpoint saws cannot be sharpened at all and get replaced.

Chisels and plane irons get honed flat on the back first - once, properly - then bevel-side on progressively finer stones at around twenty-five to thirty degrees. A honing guide is worth its small cost.

Chapter 13: Handles, Rust, and Storage

Handles

Wood handles fail at the head and at the neck just below it, and both failures are caused by neglect.

Check for cracks, splinters, and looseness before use. A loose head on an axe or a maul is a genuine danger, since the head can fly off in the middle of a swing.

Dry air shrinks a handle and loosens the head. Soaking the head end in water swells it back temporarily and is a stopgap, not a repair - the wood shrinks again and the fit gets worse each time. The real fix is to re-wedge it or replace the handle.

To replace a handle: drive out or drill out the old wedge, punch the remains of the handle out of the head, fit the new handle by paring with a rasp until it seats fully, saw the kerf if there is not one, drive the wooden wedge in with glue, cut it flush, and add a metal wedge across it. Take your time on the fit; a handle that seats properly does not loosen.

Sand new handles smooth and finish them with boiled linseed oil rather than varnish. Varnish looks good, then cracks and raises blisters. Linseed oil soaks in, keeps the wood from drying out, and can be reapplied any time. A wipe once a year is enough.

Fiberglass and composite handles do not need any of this, and they cannot be repaired when they fail. Both approaches are defensible.

Rust

Rust is water plus air plus time, and controlling water is the whole game.

Dry tools before storing them. Wipe them off. This is ninety percent of rust prevention and it takes ten seconds.

Oil them. A wipe of light machine oil, a spray of a moisture-displacing product, or a wipe with an oily rag kept in a can by the door. For garden tools, the traditional bucket of sand with a little oil worked through it does the job - plunge the shovel in a few times on the way past.

Control the humidity in the shop. An unheated shed swings through the dew point constantly, and condensation forms on cold metal every time warm air moves in. A small dehumidifier, or a light bulb in an enclosed cabinet, or desiccant in a toolbox, all help.

Removing existing rust: light surface rust comes off with steel wool or a scouring pad and oil. Heavier rust responds to a wire wheel, or to soaking in white vinegar overnight followed by a thorough scrub, rinse, dry, and immediate oiling - vinegar leaves bare metal that will flash-rust within hours if you do not oil it right away. Deep pitting is permanent.

Storage

Hang tools rather than piling them. Piled tools damage each other's edges and rust where they touch.

A shadow board - outlines painted on plywood - sounds fussy and it works, because you can see at a glance what is missing before you lose it in the field.

Cover cutting edges. Sheaths for axes and machetes, a bar cover for a chainsaw, a scrap of hose split lengthwise over a saw tooth line.

Store batteries out of extremes. Lithium tool batteries lose capacity permanently in heat and refuse to charge when very cold. Somewhere in the house is better than the shed, and storing them at partial charge is better than full or empty for long periods.

Store fuel and oil separately from tools, in a cool dark place, and check your local fire code on quantities.

Chapter 14: The Shop Itself

What is actually worth owning

Wrenches and sockets in both metric and SAE. Most modern small engines are metric; older American equipment is not, and you will need both no matter what anyone tells you. A 3/8 drive socket set covers nearly everything, with a 1/4 drive for small fasteners and a 1/2 drive with a breaker bar for blade bolts.

Screwdrivers including the flat blades wide enough for the slotted screws on carburetors, which cam out and get chewed up by anything too narrow.

Pliers: slip-joint, needle-nose, and locking pliers.

A spark plug socket in the sizes your equipment uses, and a feeler gauge for gapping.

A torque wrench. Aluminum small engine castings strip easily and "good and tight" is not a specification. A moderate 3/8 drive click wrench is inexpensive and prevents a class of expensive mistakes.

An oil extractor pump. Twenty or thirty dollars and it changes small engine oil changes from a wrestling match into a two-minute job.

A funnel set, drain pan, shop rags, and a wire brush.

Files: a bastard mill file for axes and blades, round chainsaw files in your chain's size, a flat file for depth gauges.

A bench vise. The single most useful thing you can bolt to a bench.

Penetrating oil, and the patience to apply it and walk away for an hour. Nothing you can do with a wrench is as effective as time.

An inline spark tester and a compression gauge, both cheap, both turn guessing into knowing.

A multimeter for the electrical side of anything with a battery.

Fasteners and consumables

An assortment of bolts, nuts, washers, and cotter pins in the common sizes, sorted in a small parts cabinet, will save more trips to town than anything else in the shop.

Keep on hand: engine oil in the viscosity your equipment takes, two-stroke oil, bar and chain oil, grease and a grease gun, fuel stabilizer, thread locker, anti-seize, and a can of penetrating oil.

Keep spare spark plugs and air filters for your main machines. They cost very little and they are what you need at seven on a Sunday morning.

Working safely

Eye protection for anything spinning, cutting, grinding, or under spring tension.

Fire. An ABC extinguisher mounted by the door, so you pass it on the way out rather than having to go past the fire to reach it. Rags soaked with linseed oil can spontaneously combust as the oil cures - spread them flat outdoors to dry, or keep them in a sealed metal can.

Ventilation whenever you use solvent, carburetor cleaner, or run an engine, and never run an engine in the shop, period.

Lighting. Most shop injuries happen because somebody could not see what they were doing.

Batteries. Lead-acid batteries produce hydrogen and can explode; charge them ventilated, and connect the ground clamp last and away from the battery. Battery acid needs eye protection.

Lifting. Get help or use a hoist. A riding mower deck is heavier than it looks and it has edges.

Battery equipment

A growing share of the tools on an ordinary place now run on lithium batteries - drills, chainsaws, mowers, trimmers, blowers. They have no fuel to go stale and no carburetor to varnish, which eliminates the leading cause of failure in this entire book. What they have instead is a battery, and the battery is the whole machine.

Pick one platform and stay on it. The batteries are the expensive part and they do not cross between brands. Two or three batteries on one system beats a shed full of orphaned chargers.

Heat is what kills lithium cells, permanently and invisibly. A battery left on the dashboard of a truck in July, or in an uninsulated shed through an Alabama summer, loses capacity it never gets back. Store them indoors at room temperature.

Cold does not damage them, but it does stop them. Most packs refuse to charge below freezing and deliver noticeably less runtime in cold weather. Bring them inside, charge them warm, and carry the spare in a coat pocket in winter.

For long storage, leave them at partial charge rather than full or empty - somewhere around half is the usual advice, and most manufacturers say the same. A pack stored full at high temperature ages fastest of all.

Do not charge a visibly damaged, swollen, or water-soaked pack, and do not throw any lithium battery in household trash or a burn barrel. Damaged cells start fires that are very difficult to put out. Most home improvement and battery retailers take them for recycling free.

A battery chainsaw is a legitimate tool for limbing, bucking small rounds, and storm cleanup, and it will not replace a gas saw for felling or a day of firewood. Book 24 covers that trade-off in more detail.

Manuals and records

Keep the manual for every machine in one binder or one folder on your computer. When you lose one, search the model number plus "manual PDF" - nearly every manufacturer posts them free.

Keep a maintenance log. Machine, date, hours, what you did. It takes fifteen seconds, and it turns "I think I changed that oil last year" into a fact. It also raises resale value, since a documented machine sells for more.

Part Five - A Year Of Maintenance

Chapter 15: A Year of Maintenance

Late winter

The best time to service everything, because the shops are slow, the parts are in stock, and nothing is urgent.

Service the mower before the grass grows: oil, air filter, plug, blade sharpened and balanced, deck scraped. Service the tiller. Sharpen and oil all the hand tools. Check handles. Order the parts you meant to order in October.

Tap maple if you do that, which is a Book 26 topic.

Spring

First mow of the season, and the first run of every machine that sat. Expect the ones you did not store properly to give you trouble, and note which ones, because that is your own report card.

Start the string trimmer with fresh mix. Test the generator under load. Check the chainsaw over before spring storm season.

Summer

Ongoing: check oil before every use, sharpen the mower blade a couple more times, keep the air filters clean in dusty conditions, and keep an eye on fuel age in the cans.

Mid-season oil change on the mower if you are running a lot of hours.

Fall

The big one. Firewood season means saw maintenance - new chain or a full sharpening, bar flipped and cleaned, air filter, plug, fuel and bar oil topped off.

Test the generator seriously now, before storm season, under real load, and buy the oil and filters you would need during a week-long outage.

Winterize what is done for the year: pressure washer pump saver, pump drained, tiller tines cleaned and oiled, mower stored per Chapter 2.

Winter

Snow equipment gets the attention if you have it. Otherwise this is bench time - sharpening, handle work, organizing fasteners, reading manuals, and fixing the things you worked around all summer.

What you will get wrong

You will leave fuel in something. You will find out in April.

You will overtighten a bolt into aluminum and strip it, and you will learn what a helicoil is.

You will spend an hour on a no-start before checking whether the switch was on.

You will buy a cheap tool, break it, and buy the good one, having paid for both.

None of that matters much. The mistakes that matter are the ones in the safety notice, and those are avoidable by simply never taking the shortcut.

What good looks like

Everything starts. Not usually - every time, on the second or third pull, in April after sitting all winter, because the fuel in it is either fresh or absent.

You know how many hours are on each machine. The blade is sharp, the chain throws chips instead of dust, and when something does quit you find the cause in ten minutes instead of loading it into the truck.

That is not mechanical talent. It is fuel discipline, a two-minute look before you start anything, and writing down what you did.

Quick Reference

The three killers

1. Stale fuel. Sixty days maximum without stabilizer. 2. Dirty air filter. 3. Old, low, or absent oil.

Two-stroke mix

Ratio	Per gallon	Per 2 gal	Per 2.5 gal	Per 5 gal
50:1	2.6 oz	5.1 oz	6.4 oz	12.8 oz
40:1	3.2 oz	6.4 oz	8.0 oz	16.0 oz
32:1	4.0 oz	8.0 oz	10.0 oz	20.0 oz
25:1	5.1 oz	10.2 oz	12.8 oz	25.6 oz

Oil rated ISO-L-EGD or JASO FD. When in doubt, mix richer. Mixed fuel: 30 days, 60 absolute maximum.

Four-stroke oil

Oil	Conditions
SAE 30	Warm; hard starting below 40 degrees F
10W-30	Varying; may raise consumption above 80 degrees F
Synthetic 5W-30	All temperatures
15W-50	Continuous heavy duty

Walk-behind: about 15 - 18 oz, change every 50 hours or annually. Rider: about 48 - 64 oz, every 100 hours or annually. First change at 5 hours on a new engine. No additives.

Fuel rules

E10 maximum. E15 is prohibited by federal law in small nonroad engines - sometimes sold as "Unleaded 88." Date every can. Stabilize at the pump. Store equipment either run dry or full and stabilized, never in between.

No-start checklist

1. Switch on, fuel present, valve open, safety interlocks satisfied, choke correct 2. Spark - strong blue at the gap 3. Fuel - is the plug wet after several pulls 4. Compression - clear resistance on a slow pull

Starting fluid firing briefly then dying means the problem is fuel delivery.

Symptom guide

Symptom	**Usual cause**
Runs on choke only, dies at idle	Plugged idle circuit
Surging idle	Partly plugged idle circuit
Dies hot, restarts cold	Ignition coil, or blocked tank vent
No spark after hitting an object	Sheared flywheel key
Power loss under load	Fuel restriction, plugged spark arrestor, dull blade or chain
Smokes, oil out the breather	Overfilled crankcase

Chainsaw

Sharpen every tank, and immediately after touching dirt. File size, top plate angle, and depth gauge setting come from your chain package or manual. Equal strokes on every cutter. Check depth gauges every few sharpenings.

Flip the bar at each chain change. Clean the groove and oiler hole. Bar and chain oil only - never used motor oil. Roughly one tank of bar oil per tank of fuel.

Tension cold: snug to the bar, turns by hand. Re-check after the first minutes of cutting. Never tension hot.

Test the chain brake before every use.

Mower blade

Follow the existing bevel, about 30 degrees . Butter-knife sharp, not razor sharp. Balance it. Mark the down side before removal. Keep it cool while grinding.

Generator

Outdoors only, 20 ft from the house, exhaust pointed away. CO alarms inside. Never backfeed a wall outlet. Transfer switch or direct cords only.

Watts = volts x amps. A 5,000 W generator at 120 V delivers about 41 A.

Size for total running watts plus the largest single starting surge, plus margin.

Exercise monthly for 20 - 30 minutes under load. Store dry or stabilized. Keep oil, filters, plugs, and fuel line on hand.

Refuel with the engine off and cool.

Sharpening angles

Tool	Approach
Axe	Convex, follow existing bevel, do not thin
Maul	Blunter than an axe
Shovel, hoe	About 45 degrees , working edge only
Bypass pruner	Outside bevel of the cutting blade only
Knife	About 20 degrees per side, raise a burr, then remove it
Chisel, plane iron	Back flat once, then 25 - 30 degrees on the bevel
Mower blade	About 30 degrees , deliberately dull

Glossary

Backfeeding - Energizing house wiring from a generator through an outlet, which sends power onto the utility lines. Dangerous, illegal, and never acceptable.

Bar oil - Purpose-made tacky oil for chainsaw bars and chains. Not interchangeable with motor oil.

Choke - A plate that restricts incoming air to enrich the mixture for a cold start.

Compression release - A camshaft feature that bleeds cylinder pressure at cranking speed to make starting easier, and which makes measured cranking compression read lower.

Depth gauge (raker) - The tab ahead of each chainsaw cutter that limits how deep it bites.

Flywheel key - A soft metal key indexing the flywheel to the crankshaft. Shears deliberately when the blade strikes an object, throwing off ignition timing.

Governor - The mechanism that holds engine speed constant as load changes.

Idle circuit - The small passages that feed fuel at low throttle. First thing to plug with varnish.

Jet - A calibrated orifice metering fuel into the airstream. Clean it; never drill it.

JASO FD / ISO-L-EGD - Current performance specifications for air-cooled two-stroke oil.

Phase separation - Ethanol and absorbed water separating out of gasoline into a corrosive layer.

Spark arrestor - A screen in the muffler that stops hot carbon from leaving the exhaust. Required on public land in many places, and it clogs.

Starting watts - The brief surge an electric motor draws at startup, often two to three times its running draw.

Transfer switch - A device that safely isolates house circuits from the utility before connecting a generator. Installed by a licensed electrician.

Varnish - The hard residue left when gasoline evaporates. The cause of most carburetor failures.

Where To Get Help

Your equipment manufacturer. Manuals, parts diagrams, and specifications are posted free by model number on nearly every manufacturer's site. The model and serial numbers are on a tag or sticker on the engine and on the frame, and the two are usually different.

Your local small engine shop. Worth cultivating. They will tell you what is worth fixing and what is not, they stock the parts, and a shop that knows you will fit you in during the busy season.

Cooperative Extension. Many state extension services publish fact sheets on fuel, ethanol, and equipment maintenance. Oklahoma State's work on ethanol blends in small engines is the source for Chapter 2's storage guidance.

EPA. Fuel regulations, including the prohibition on E15 in nonroad engines, and the required pump labeling.

A licensed electrician. For any generator connection to house wiring. This is not a place to save money.

Manufacturer service schools and dealer training. Some dealers run owner clinics before the season. Free, and worth an evening.

About The Author

Hudson Briggs writes practical guides for people learning to do things themselves. His work covers gardening, livestock, food preservation, homestead infrastructure, and the ordinary skills that keep a small place running. He believes in fuel discipline, in a two-minute look before you pull the rope, and in writing down what you did so you do not have to remember it.

www.ingramcontent.com/pod-product-compliance
Lightning Source LLC
LaVergne TN
LVHW010739061126
851857LV00014B/1551
* 9 7 9 8 2 4 0 8 4 2 9 9 3 *